Influence of Foot-Length on the Gait of Horses

An over-view of tests plus details of purpose and the background reasoning behind arranging and conducting these tests by Lorry Wagner, followed by the below report.

Dr. Auer's report detailing the series of force-plate recordings showing the difference of location of peak-force impact between long and short hoofs of two Arabian stallions.

Conducted at the Texas A&M University, College Station, Texas
Texas Veterinary Medical Center
College of Veterinary Medicine
Department of Large Animal Medicine & Surgery
J. A. Auer, Dr. Med. Vet, MS
Associate Professor, VLAM

Cover
Golden Horse Heads
by Stacey Mayer

ISBN: 978-0-578-13033-0

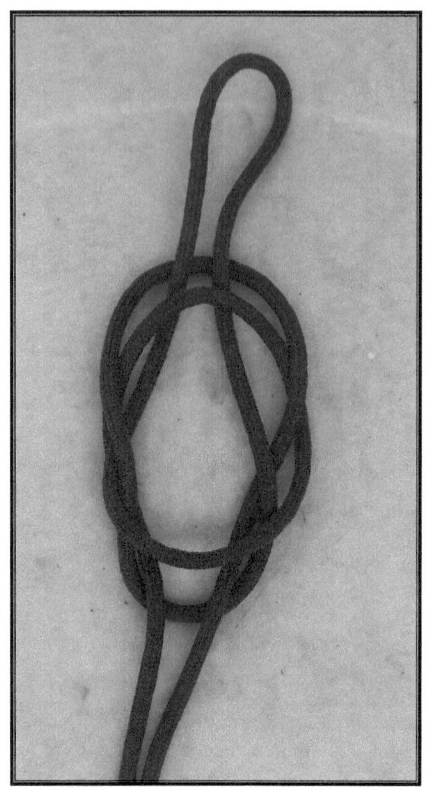

Pictured above is the knot which is configured to attach a fiador to the bosal heel knot of the hackamore.

The author of this book learned the traditional Vaquero (sometimes called Early California) training including the use of the hackamore. Throughout this book various pictures are included showing this traditional horsemanship.

TABLE OF CONTENTS

Page No.

Pictures of Examples of Vaquero Equipment, the Author and Her Horses

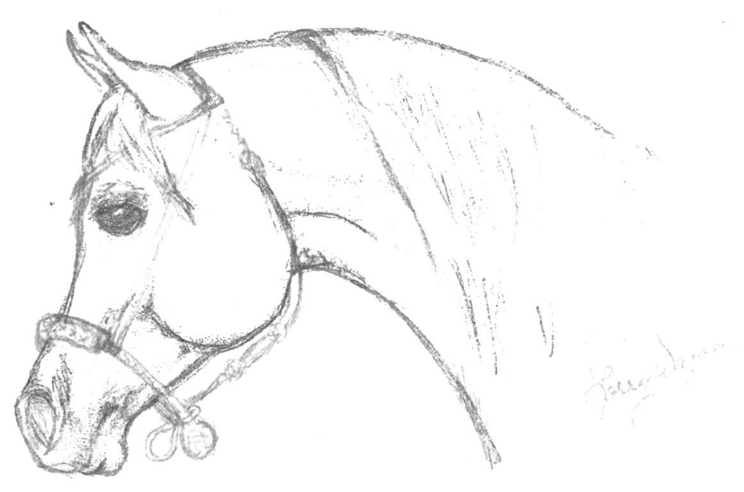

Pictured above is the knot attaching the fiador to the bosal of the hackamore. The rope hanging in the back continues to configure the rest of the headstall which keeps the hackamore from slipping forward over the ears and off the horse for any reason.

Reins shown in the background are braided calf strings in 16 plait braided by the author's teacher and step father,

William "Wink" Chappell

The Texas Veterinary Medical Center
College of Veterinary Medicine
Texas A&M University

Department of
Large Animal Medicine & Surgery

November 20, 1987

Mrs. Lorry Wagner
Sierra Dawn Arabians
8222 Athel Ave.
Inyokern, CA 93527

Dear Mrs. Wagner:

Thank you for your letter of November 2 and the additional information you sent me. I am very impressed by the way you approach this problem and especially the letter to your committee members was to the point. Briefly I would like to explain, as you asked, what we would evaluate with the horses.

I preferably like to have Arabian horses with, as suggested, 5-5 1/2" of hoofwall and standard shoeing. These horses would be walked and jogged over our gait analysis system which is briefly described in the enclosed brochure. We would then go ahead and trim the horse back to the normal hoof lengths which would be 3 1/2" and again, the horse will be walked over. The patterns which we will get from these two evaluation sessions of the same horses with different hoof lengths will be compared. We will compare the time the animal was on the ground with each foot and stride length as well as the detailed parameters as each foot was placed and the horse's weight was moved over that foot. We are particularly interested in finding out how much time the animal spent in the position just before lifting off the toe.

I am positive that we can find significant differences similar to those in race horses with toe grabbs. Parallel we would film those horses with high speed videography and analyze the motions with the help of the orthogram given from the gait analysis. I would suggest we do one or two horses first and when we find the differences which we expect, we can add an additional number of horses through the study to convince your board members and hopefully the community of Arabian horse owners that shoeing as proposed (5-5 1/2" hoof length) would be detrimental to the horse.

I hope that this information is sufficient for you to convince possibly Mr. Wayne Newton to help study the problem. If I can be of any additional help, please do not hesitate to contact me at any time.

Sincerely,

J.A. Auer, Dr. Med. Vet., MS
Associate Professor, VLAM

JAA:nb

College Station, Texas 77843-4475 • 409-845-3541

The Texas Veterinary Medical Center
College of Veterinary Medicine
Texas A&M University

Department of
Large Animal Medicine & Surgery

August 16, 1988

Ms. Lorry Wagner
Sierra Dawn Arabians
8222 Athel Avenue
Inyokern, California 93527

Dear Ms. Wagner:

Enclosed I am sending you the report of the study we performed on the two Arabian Horses owned by Ms. Chapman. I hope that the results and the figures convince you that you were right in assuming that long toes are detrimental to the movement motion in the Arabian Horse. We came to the same conclusion and were able to show that normal foot length and angle decrease peak force borne on the forelimbs compared to the rear limbs. The majority of the weight should be borne on the rear limbs with a rider on the horse. I also enclose the bill for the study with this report. I hope that you are satisfied with what we have found and if I can be of any additional help please do not hesitate to contact me anytime.

Sincerely,

J.A. Auer, Dr. Med. Vet, MS
Professor, VLAM

JAA/agm

Enclosures

College Station, Texas 77843-4475 • 409-845-3541

The Texas Veterinary Medical Center
College of Veterinary Medicine
Texas A&M University

Department of
Large Animal Medicine & Surgery

September 28, 1988

Lorry Wagner
Sierra Dawn Arabians
8222 Athel Avenue
Inyokern, California 93527

Dear Ms. Wagner:

Enclosed I am returning your first draft of Step One in the Fight to Preserve the Bedouins Arabians. I think you did a very good job and I am impressed by the through evaluation you have done on my report. You get the essence out of it and cited the number correctly, also you interpreted them correctly. I made a few changes to clear up some misunderstandings but I did not have to do hardly anything on your report. I wish you great success with your project. If I can be of any additional help, please do not hesitate to contact me any time.

Yours sincerely,

J.A. Auer, Dr. Med Vet, MS
Professor

JAA/er

Enclosures

College Station, Texas 77843-4475 • 409-845-3541

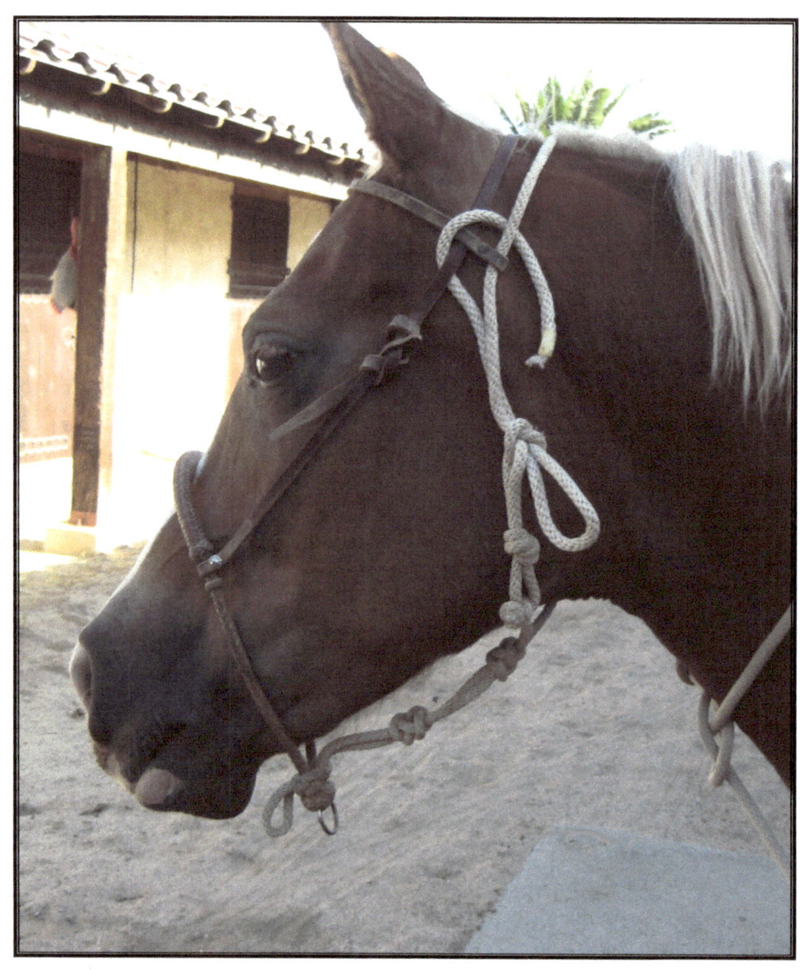

Leit Motif, Arabian Mare, wearing a light bosal showing the fiador in conjunction with the headstall completing the hackamore as traditionally designed and configured Vaquero style.

Note the "quick release" tie where the part comes over behind the ears and attaches to the line that comes from the split under the throatlatch. Vaquero traditional design for their equipment almost always used a "quick" release that firmly held when tied.

Preface

Most of the contents of this book were written at the time the force-plate tests were conducted and specifically address the Arabian horse breed; however, these test results are very applicable to any horse of any breed, including grade horses.

I will first state that the force-plate tests reported within this book were conducted because of the problems in the Arabian horse breed's show-ring trends of the day. Some of the supporting information I share for conducting this testing project is specific to the Arabian horse breed in the area of that breed's history, natural movement and my own background for purpose of justifying my pursuit of putting the soundness, health and well being of the horse as the primary charge of owners and riders above all else. However, some of this "Arabian horse" information is applicable to all breeds and the actual test results are applicable to all horses in maintaining sound and healthy horses, no matter what that breed's natural movement may be.

All horses must have proper foot care focused on making sure the naturalness of movement of that particular horse is not compromised. Otherwise, improper care can lead to lameness and other problems in many places over the body. There are 4 graphs within this report recorded by high-speed cinematography showing how the entire body carriage is affected by long versus short feet which shifts the peak concussion point of impact within the foot. These tests are just a start and do not even begin to address the problems of pads, weight of shoes and many more man-made creations installed on horses feet in the name of "better movement." This latter is basically secondary to the focus of these tests, but are very prevalent and most often done in ignorance of how "artificiality" impacts the natural functions of the hoof which reverberates up through the body.

It seems over the years most of our Arabian horse breed's show-ring competition has led to very interesting creations nailed (or glued or banded) onto the foot which totally destroy natural movement resulting in creating lameness and many other problems; however, those interested in other breeds can find the parallels for their breed very easily, some which were imposed on horses long before the Arabian breed turned away from natural movement.

I share a lot of the history of the Arabian horse within the following pages, since that history is so important to the question of "do we preserve the original Arabian horse or modernize our breed to the whims and fancies of today's generations?" This "modernizing" is a primary cause of many of the "creations" we now see on our horses' feet. The history of each breed is important to this subject, since each breed has it's own history and can be substituted in relationship to the results of the testing reported herein.

I share my own background justifying my knowledge on this subject and why I feel so strongly about exposing the injustice done to not only the Arabian breed but all horses. I have also included a few pictures sharing bits of Vaquero traditional equipment and horses trained by these methods that might be of interest. The good health and soundness of all horses is imperative to maintain and should be far above interfering with the natural horse in priorities. All owners must endeavor to understand what is needed to make sure their horses are healthy, sound and happy.

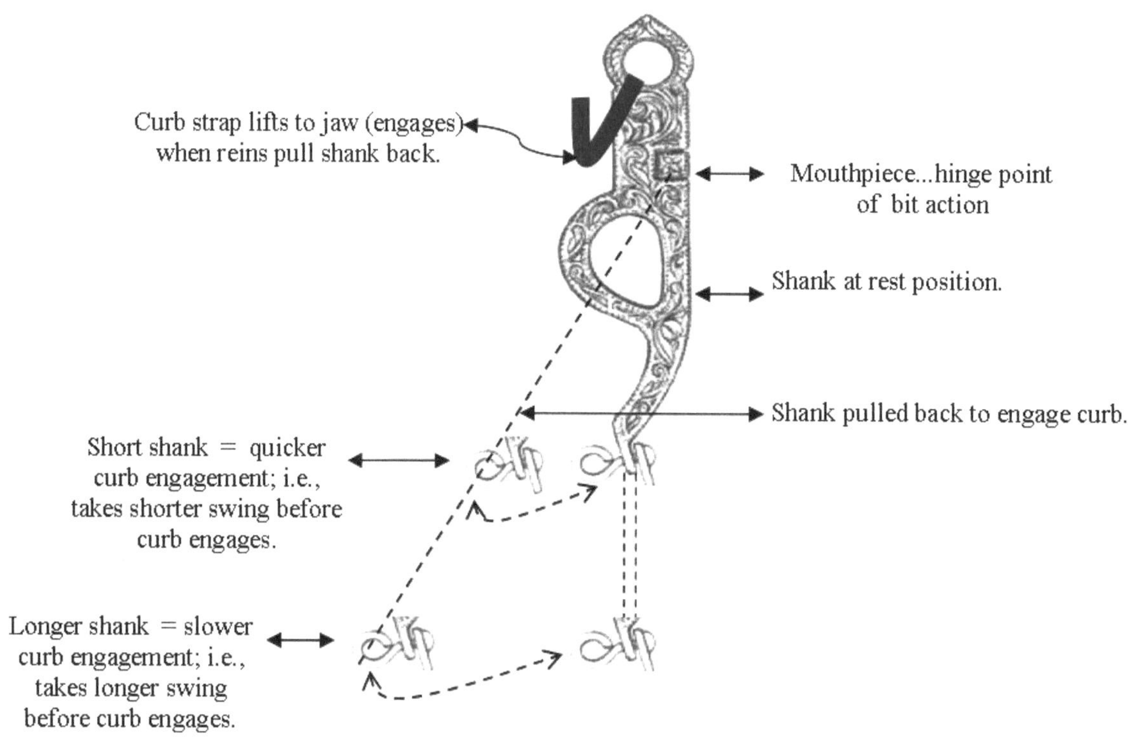

The above drawing shows how the leverage of a curb bit really works in coordination between the curb and the rein connection at the bottom of the shank when communicating with the horse.

The Vaquero tradition included training horses to responding to very finite communication between the rider's hands on the reins in coordination with the rider's leg aids which start with basic training in the snaffle and/or hackamore.

Influence of Foot Length on the Gait in Horses

Expanded from the Original Testing Report Which Specifically Addressed the Arabian Horse to Embrace All Horses

By Lorry Wagner

In the Beginning

I begin this book by addressing a lot of historic background addressing how the Arabian horse evolved through many centuries which is why I feel so strongly that preservation of that original Arabian horse is vitally important to all horses on into perpetuity. As this subject leads to the reasons that inspired me to having these tests conducted, it will be found that many of the same problems have led many other of our light-horse riding breeds down the same path, thus making these test results applicable to all horses. I include extensive background on myself to share how my life with horses sent me down this path to organize this testing.

After having bred, raised, trained and participated in competitions, both the endurance trail and show-ring arenas, in addition to working cattle in the high meadows of the Sierra Nevada Mountains and pleasure riding in the Mojave Desert of California with my Arabian horses for nearly 20 years, my experiences and observations led me to jump into the committees and politics of the International Arabian Horse Association in 1978. I saw so much happening that was detrimental to my breed I decided my concerns must be addressed from a position with more impact that just a member voicing an opinion and frustration at local meetings. I also wanted to share the very unique educational opportunities I had been so fortunate to have received with so many others that would never have that opportunity. I had joined the closest Arabian Club that was 120 miles away not long after I bought my first Arabian and drove to the meetings once a month. I then organized my local Arabian Club in 1976. Throughout the passing of those years I became aware of the direction things were going. I moved up to Secretary of Region 2 of the International Arabian Horse Association (IAHA) in 1978 which brought me to working at the National level of IAHA. I was elected to the IAHA Judges' Seminar Commission in 1986 and the position of Region 2 Director in 1988. In the meantime, I also became a life member of WAHO in 1982 and became deeply involved in this organization by 1988 which brought me to traveling for the Arabian horse to many countries around the world.

With that summary of my early days with Arabian horses, I go back and address a lot of historic background and how the Arabian horse evolved through many centuries which is why I feel so strongly that preservation of that original Arabian horse is vitally important to all horses on into perpetuity. As this subject leads to identifying the reasons that inspired me to have these tests conducted, it will be found that many of the same problems have led many of our light-horse riding breeds down the same path, thus making these test results applicable to all horses.

When I started breeding Arabian horses in 1960, the ultimate goals of all Arabian horse breeding programs and/or show strings were to emulate, duplicate, and preserve the horse the Bedouin bred and rode to war, to hunt, to race, and/or to participate in fun and games as was their way of life for century after century. From the first Arabian horses imported to America sometime in the 1700's up through the years until the late 1960's the breeders and owners of this ancient breed took great pride in and accepted responsibility for preserving and perpetuating the magnificently-balanced, gracefully-moving, energy-efficient, high-density and flat boned, strong-sinewed, iron-muscled, dependably-SOUND, and tough desert horse with his kind, loving, and superbly intelligent mind. That all these necessary attributes were combined in a beautiful-to-behold package is a tribute to the many centuries of Mother Nature's superb artistry...what we today term "type"... created in the environment of the Middle Eastern deserts in partnership with the Bedouins who lived there. There are still some of us that are old enough to remember when we DEMANDED that our Arabian horses be shown stressing the importance of NATURAL appearance; NATURAL long, low, floating trots; NATURAL, NATURAL, NATURAL; and above all...SOUND which kept us on the path to the above goals!

The Arabian horse developed these unique characteristics through all those centuries of surviving in one of the most harsh environments on earth and being used by the Bedouins in some of the hardest usage man ever asked of any equine by incorporating selective breeding by those Bedouins for these very important characteristics that were necessary for the survival of both horse and himself. Weakness was bred out by the natural survival-of-the-fittest environment which, over the centuries, created a concentrated and extremely strong gene pool. It is well known that these characteristics are the concrete foundation that make the Arabian horse unique and different from all other breeds. The historic creation and preservation of the Arabian is also the reason why all of today's light-horse riding breeds trace back to that Arabian gene pool from which sprang each breed's genetic foundation, or in simple language, all these breeds trace their creation to the Arabian horse's very strong and unique gene pool. Each of these light-horse riding breeds were created by selective breeding drawing emphasis on individual attributes needed at that time which evolved into the different breeds, each with their own unique specialties.

At this point, I believe it is appropriate to justify my own credibility for not only having these tests conducted, but also for explaining in great detail from where I draw the conclusions expressed herein; i.e., from my actual experience.

I fell in love with the Arabian horse when I received "The Big Book of Horses" from my grandmother for Christmas when I was 5 years old. The first page was a double-page spread of a picture of a gorgeous white Arabian. From that day on I searched the local Library for every book on horses I could find which, of course, included the Black Stallion and Red Stallion series by Walter Farley and "Smokey," among many others. After I graduated, found a job and saved a bit of money, I bought my first Arabian, Ibn Nasah from Foster Smith of Lone Pine, California, and went back and bought the stallion, Rasia, the next year, also from Foster. Rasia was a hit and I was suddenly breeding many local mares. Thus, Sierra Dawn Arabians evolved. In the early 1960's I had the wonderful opportunity to learn to train correctly. My first instructor was Ed Hale who helped me train my gelding, Ibn Nasah. Ed was a cowboy and terrific horseman originally from Arizona who had ridden a green horse in training leading a pack horse from Arizona to Southern California years before.

Then William A. Chappell (known to many as "Wink") came into my life and we became partners in conducting a public training stable in conjunction with Sierra Dawn Arabians. It started with the agreement that I would provide the stabling and care and Wink would teach me the Early California Vaquero tradition of training. After apprenticing for 5 years under Wink's thumb and being fully schooled in the "classic" tradition of the Early California Vaquero methods, Wink and I continued that public training stable in partnership until he retired from training for the public but continued training my Sierra Dawn Arabians, while I continued with taking public horses for a few more years in addition to splitting the Arabians I bred with Wink when they reached the age for training.

Wink grew up on the Quarter Circle H ranch owned by Roland Hill located in the Tehachapi, California, area. Roland not only raised cattle, but also Morgan horses...he had about 200 brood mares which were selectively grouped and held in fenced sections of 50 or so mares and one stallion who was also selectively chosen for that group. The resultant foals were started at 3 to 4 years old, some for cattle work and quite a few were shipped to New York for the police force. Wink rode these horses from the time his father (who was also a trainer in the Vaquero tradition) would sit him on a horse before he could really walk. When he was about 12 years old, the tradition of the time was to put those young would-be trainers on spoiled and rank horses that were tough to deal with. Cowboys of that day were the same as today...some could ride well and some made a mess of a good horse and that horse would be sent to the "condemned" field where it would be picked up by the trainers to re-condition for the next year. Some of these horses were pretty ornery and "the boys" got lots of practice. The reasoning behind this was to get the "hot rod" out of those boys. The ones that stayed with it learned to ride anything and when they "graduated" to the young 3 and 4 year olds Wink's description was "it was like taking candy from a baby" to train them.

He then moved on when he was in his late teens and early 20's to finish his education under those great Vaqueros of that day to become the wonderful horseman he was. He "apprenticed" for approximately 10 years starting his first 2 years learning the snaffle bit, 2 years learning the hackamore, 2 years learning the half-breed and the last 4 years perfecting knowledge of the spade bit. Each of his trainers were the best experts in that particular phase of training, but all were accomplished in all areas. All these years he also added to his "cowboy" knowledge the Vaquero's perfection of proper care of the feet to keep one's horse sound, including doing the trimming and shoeing one one's own horses as to that horse's individual need. That was simply a requirement of the times.

To tie this background into the subject of this book, part of my training also included Wink's teaching me to know and understand how horses feet should be managed. I never saw a horse with a foot problem that Wink could not correct with simple farrier work. We shod our own horses and most of the horses in training with the exception of those whose owners used their own farriers; however, we over saw them and made sure no problems were created. Thus, I learned to 1) ride the horse and feel his movement, 2) determine if there were a problem in that movement, 3) if yes, go back to the barn, dismount, unsaddle, pull shoes, re-trim and reset the shoes...yes, me bending my back and 4) re-saddle and ride and feel that horse again. It was a marvelous education and I learned how important it is to ensure well-managed feet to enable each horse to reach their ultimate correct movement and balance. This part of my learning to train gave me a very in-depth background to pursue this testing project. Wink was not only a great trainer, he was also the best farrier I have ever known. I have taught many young fellows some of the finite "tricks" to keep feet correct over the years after Wink passed away and I could no longer physically bend over to trim and shoe my horses.

My dad had died in 1961. Mom met Wink when he came into my life and they married in 1969. Thus Wink was stuck with me for the rest of his life as his step-daughter. We continued to train horses together for a full 30 years, thus my education had a wonderful exposure to over 200 horses of learning all those years. When Wink was about 72 he quit starting horses under saddle, but I would start them and he would finish them into the bridle. Wink rode until he was 82 and started having heart problems. He passed away 2 hours before his 85th birthday in 1994. I still miss him terribly. He was one of the last of the great trainers who learned under some of the finest Vaquero trainers of the early 1900's. Horses trained by this method are the western equivalent to the finest finished Classic dressage horse.

We rarely finished a client's horse in the spade bit, since this training takes 2 to 3 years to accomplish and most riders are not trained to handle the spade. We bridled client's horses with the bit each horse accepted and liked which went from low ports to medium ports and a variety of shanks which worked well with that particular mouthpiece and horse.

I will also state here to clarify for those who may not be aware, the Vaquero training evolved in different scales of refinement, depending upon what part of California this was tradition and the kind of horses that were in that area. I refer to "classic" tradition that I learned which was practiced in mid-California from Tehachapi westward to Santa Barbara area. The Tejon Ranch located southeast of Bakersfield, California, (who had imported a number of Arabians from the Middle East sometime in the late 1920's) was established in 1850 as part of an old Spanish land grant and employed some of the finest Vaquero horsemen of the early 1900's who produced finely trained horses. Many of these Vaqueros are the ones who trained Wink.

Horses trained by this method are the western equivalent to the finest finished Classic dressage horse. The Spanish Ranchos established in California in the 1600's brought not only their horses of Arabian descent (as a result of the Moor's invasion riding their Arabian horses in Spain's past history) but also their horsemanship which was adapted into working cattle on their vast rancheros. This evolved into what I call the "classic" tradition which was influenced by training Arabians, Morgans and Thoroughbreds (and half-bred offspring of same which were popular for ranch work) which are finely bred and much more sensitive and intelligent than the cold-blooded grade horses (often thick skinned and very stubborn with lots of fight in them) which came later and were prevalent in mostly the more northern areas of California. Those horses required much rougher handling; however, the "basics" of the Vaqueros' traditional methods were still there...these tough horses just required a tougher approach to get their attention.

I learned the tradition one uses for the finely bred horses which produces those wonderful well-bridled horses that equate to the true Classic Dressage traditions with the major difference being the Vaquero training is based upon the training of the original war horse but was adapted to the requirements of the horses for working on cattle ranches and Classic Dressage which is also based on maneuvers required by history's war horses but is now adapted to modern-day usage. Both traditions have the same roots back in centuries when a well-trained war horse had to be instantly responsive to leg aids while its rider handled weapons defending them both which was an absolutely necessary team coordination in order to have the chance to survive when engaged in battle. Having explained my background and from where my cause comes and why I chose the Arabian as my horse which led to my organizing these tests, here's a bit more.

I have bred, raised, and trained many, many Arabians since 1960. For years I showed both my own horses and client's horses in nearly all disciplines in the show ring plus a bit of endurance competition and kept my horses sound. The same is true during all the times of working cattle taking a horseman's holiday doing trail and pleasure riding both through the Sierras and the Mojave desert. At times I would be more than 50 miles from another living soul. Strength, intelligence, and especially SOUNDNESS is a must in these last circumstances. IT'S A LONG WALK HOME! I have never had to take that walk.

I strongly maintain the natural movement in the various gaits that the Arabian executes (meaning in the natural manner he moved when he came directly from the Bedouins) are correct for maintaining the health, soundness, and preservation of this breed. I never had a horse I rode and managed the feet go lame and found all breeds, whatever their different attributes from breed to breed, must have correct foot care to stay sound. We took pride in both our own and our client's horses remaining sound. We often had horses come in for training for whom we had to correct the trimming and shoeing before riding them.

Since the late 1960's I have watched these original attributes of the Arabian horse become dangerously threatened with extinction. I have witnessed physical changes that can be made by selective breeding, political pressure in the show ring, very poor judging (usually based on the politics of "you place me today and I place you tomorrow"...the fall out of allowing our active trainers to hold a judge's card) and ignoring the history of the breed and/or judging the Arabian by other breed standards. One of the most serious examples of this latter is that we now have Arabians trotting more like Saddlebred and Morgan breeds. This kind of trot is not in line with the prized characteristics for which Arabians were revered for centuries. To change his way of going is EITHER to make him perform in a manner for which he is not meant to be used because of his unique structure (resulting in the indescribable gaits which we often see in today's English pleasure and park classes), OR to change his conformation which results in destroying the Bedouins' original Arabian horse. Since we are specifically addressing the tests based on the trot gait, I will not go into the other disciplines that have other problems detrimental to the health of the horse, and there are many of these throughout our different breeds in competitions.

One thing I observed that was very disturbing was the many, many owners, breeders, and a few trainers of Arabians who took pride in being horsemen were slowly withdrawing from the show arenas which were slipping toward a driving need to CHANGE the naturalness of the Arabian horse. I believe this was also happening or had already happened within other breeds who also were experiencing or had already experienced these same kind of problems. In the case of our Arabians it was to move and perform like other <u>man-made</u> breeds. Some of those Arabian trainers who were dedicated to the horsemen's training world had no choice but to stay in the show ring, since their livelihood depended upon income from the more recent owners who wanted their horses trained and shown to win no matter how that job was done. Some of these trainers were good friends of mine and voiced to me their frustration at being forced to compromise their principles to make a living. The drive to make these changes seemed to be based on ego and economics. In other words--the HECK with PRESERVING AND PROTECTING THE BREED, the almighty do-anything-to-win and/or dollar comes first! In the world of Arabian horses, at the time of this "drive for glory and riches" there was a tremendous influx of new Arabian owners, trainers, and judges who were, of course, exposed only to the "new fast-lane antics."

As a consequence we now have a grave problem in perpetuating the original Arabian horse as perfected by centuries of Bedouin life. I will add here that in my travels to countries in the Middle East, I found the traditional original Arabian is still being bred and preserved.

I voiced my opposition to these changes all along the way, but as most of us are, I was too busy surviving everyday life to be of any influence in stopping the direction in which the Arabian breed was headed. In 1978 my life changed and I have since that time worked toward the goals of my convictions through organizations and horse-world politics which led to this testing.

For those who are happy and satisfied with the changing of the Arabian breed that has occurred through direct influence of show ring antics since the late 1960's, the test results and concepts expressed herein will probably not be welcomed, but must be recognized as fact. For those dedicated to keeping not only the Arabian horse, but also all horses sound and healthy by retaining a natural foot for that individual horse, I am simply reflecting your convictions.

However, I hope to enlighten those many people who have not really had the opportunity to become exposed to the facts, history, legacies, and traditions of the Arabian horse from the desert; i.e., that which made the Arabian horse the most sought after equine on earth for thousands of years. To those dedicated to other breeds I bring attention to the necessity of knowing your breed's history, legacies and naturalness to always be aware of the correct foot care based upon that breed and down to that individual horse. But, primary is bringing forth this information which is the basic law for healthy and proper foot care which is extremely important to all horses, whether grade or any breed.

Ibn Nasah, Champion Arabian Costume Horse trained by the author in Vaquero style. Ibn is shown here in the bridle after having had foundation training in the hackamore.

The Road to the Force-Plate Testing

Upon watching changes within the Arabian breed taking a direction away from the original Arabian, I started questioning why...why was the Arabian show ring drifting in this direction? Readers will find many similarities to problems within their chosen breeds. To be able to find the solution, one must first recognize the cause. I believe the following are primary areas of great negative impact in the Arabian show rings.

1. People became involved in Arabians after initial experience with other breeds and never learned the unique attributes of versatility and energy-efficient movement of the Arabian. Instead, they forced the Arabian to conform to what was familiar to them. Many other breeds had developed serious problems within their breeds and many of those people moved to the Arabian breed in order to survive in both training and judging, thus bringing their different concepts, based on another breed, to the Arabian horse show ring. For me, this highlighted the importance of understanding each individual breed's history, legacy and unique attributes for the work for which bred.

2. People were new to the horse world and became involved after the influence of change had occurred and were never exposed to the naturalness of the breed. You certainly cannot expect someone to see what is wrong if they have never seen what is right in the first place. Newcomers usually (and understandably) accept as correct whatever they are first exposed to.

3. Many Arabian horse owners, trainers, and judges of today are totally oblivious to the history and origin of the Arabian horse. This is perhaps one of the largest problems since, realizing the Arabian's uniqueness in both appearance and physical abilities is a direct result of his history, this knowledge is paramount to recognizing the characteristics most important to the breed. It is my opinion that a breeder, trainer, and/or judge must have a thorough understanding and appreciation of the Arabian's legacy in order to make a sensible and responsible analysis of any Arabian for breeding, riding, or in any class to be judged. Without this very important knowledge of history, there is no basis or foundation for knowledgeable selection.

4. Last but not least, many people have never had the opportunity to ride a well-trained desert Arabian. In addition, a great many people in today's show ring have never really had the opportunity to ride and/or work any Arabian to his potential day after day, having to rely on that horse, and having to be acutely aware of keeping him sound and healthy without veterinary or artificial means. You can't work a lame horse without doing further damage. The injury does not go away just because the horse can't feel it because someone gave him a pain killer.

Thus, many of today's riders have no experience riding other than show-ring or pleasure riding. It really takes "working and depending upon" that horse to make a genuine comparison in order to recognize the grave importance of proper riding to understand how important is the care, trimming, and shoeing to keep the animal hale and fit for the next day, or maybe even for that last 10 miles home...for example, after a hard day of working cattle, miles of endurance riding or climbing mountainous trails. Since many of today's breeders and trainers have veterinarians, etc., to get them through a class or training session, I can see why keeping the horse sound and healthy by natural methods of good breeding, good horsemanship, and good farrier work would not be as critical in their minds. The walk from the ring to the stall, then to the telephone to call the veterinarian is usually not very far. We only know to whatever we have been exposed.

Shortly after my election as a commissioner on the IAHA's Judges' Seminar Commission (renamed the IAHA Judges and Stewards Education/Evaluation Commission in 1990), I attended one of our commission meetings where the controversial subject of the proper trot for the Arabian breed was brought up for discussion. Since, the trot was at that time and still is a topic of great contention within the breed and specifically the show ring and current judging trends, this Commission should have been charged with being aware and concerned with judging movement correctly. After a rather heated exchange of views, I realized too many people associated with the breed in the show ring at that time would neither understand nor accept my (nor anyone else's) explaining the reasons the natural gaits of the Arabian are not only important to preserving the breed, but are essential to maintaining sound, healthy Arabians--especially in the show ring where we now have multitudes of lame horses that cannot seem to function without "bute," pads, etc. When I started with my Arabians it was rare that one could find a lame Arabian, even in the competitive world. The horses were trimmed short and when shod wore a shoe no heavier than 12 ounces and definitely no pads and the rest of today's contraptions.

A. Background behind organization of this testing.

1. At the IAHA's Judges Seminar Commission meeting in 1987 the subject arose addressing today's trot in the Arabian English and Park classes.

2. Lively debate evolved between the 9 members of this Commission on "natural trot of the Arabian horse as he came from the desert" versus what my fellow commissioners called "action."

3. I was challenged with the statement "how do you think you can change the trends in the ring today when it is accepted that if you do not conform you cannot win?"

4. I went home from this meeting and gave that problem a great deal of thought and finally came to the conclusion that:

    a. There is a reason the Arabian horse who, as a breed in general, naturally moves in a long, low and floating motion at the trot, stayed sound doing so and could virtually maintain this gait endlessly and effortlessly. This acknowledgement leads to the obvious fact that the Arabian's conformation dictates movement at all gaits; however, the different movement of the trot is the gait most easily seen and identified by the naked eye.
    b. The only way I could prove it would be to measure the pounds-per-square-inch (psi) concussion on the foot when it hit the ground. I knew there would

be less psi concussion from a long, low floating trot than the high knee action creating a more rounded movement interfering with natural motion of the original Arabian horse.

c. I had to get black and white, undisputable proof that would show people in a manner they could see and easily understand why the Arabian horse should travel in a manner that is natural to his conformation and that the current show-ring trends in movement are detrimental to the health and SOUNDNESS of our breed.
d. The force-plate testing was my "first" step toward proving the tremendous importance of understanding the function of horses feet and the purpose of correct foot care for maintaining sound and healthy horses.

I accepted the challenge this group threw at me. I gave deep thought to their above statement "...when if you don't do it that way you cannot win," and I was determined there had to be a way to DO SOMETHING ABOUT IT!

After much deliberation, I chose a plan of attack, so-to-speak. I pursued the avenue of using the laws of physics to prove why the natural gait of an Arabian keeps him sound and healthy, even when stressed far beyond what most of us would ever ask; i.e., travel 100 miles in less than 12 hours. I have confidence that when dedicated Arabian owners, trainers, judges, and admirers are presented with these undisputable facts, they will prove to be intelligent people who care about the breed and will join me and many others in the current fight to preserve this wonderful creation of Mother Nature in league with the Bedouin.

My first step was to see what tests had already been done and what information was available. I contacted a close friend and veterinarian I have know, respected, and admired since 1964--Dr. Joe Hird of California. Dr. Hird had been very closely involved with the Thoroughbred racing industry since long before I knew him. "Doc Hird" was also very familiar with the Arabian's contribution in founding the Thoroughbred breed. He spent many years in charge of the brood mares and racing stock owned by Don Warner of Warner Bros. Studios in Hollywood. Dr. Hird listened to my theory, agreed, was not aware of any information specifically in the area of my pursuit, and suggested I contact Dr. Rooney of the University of Kentucky. I proceeded to contact Dr. Rooney who has universal recognition as one of the world's leading authorities on lameness in horses. Dr. Rooney heard me out, agreed with my theory, said "the first thing we must do is shorten the hoof," and recommended I contact Dr. J. A. Auer at Texas A & M University to establish force-plate tests of the Kaegi Gaitanalysis system. The only such piece of equipment in the United States was located at Texas A & M in College Station, Texas, at the time of this testing.

Dr. Auer received my call and was enthusiastic about the project. He too, said "the first thing we must do is shorten the hoof." My ultimate goal was to measure differences between the long, low, floating natural stride of the Arabian breed and the high-forearm, round, and shorter stride as seen in today's Arabian English-pleasure and park-class trots and all the various influences such movements have throughout the horse's body on strain and stress points within the skeletal and muscular structures, balance, and freedom of movement. However, I recognized I must find a starting place and these force-plate tests fit the requirement.

I conceded to the expertise of these two men with their knowledge and background in the area of lameness in horses. So again, we are back to one step at a time. I proceeded to the next challenge which was to raise the money needed to pay for these tests, an area I am not fond of pursuing. I did manage to raise about half of what was needed, so I just plunged forward and paid the balance, myself. Thus, by April of 1988 this testing was set up for June 27 at Texas A & M to measure the differences between a long hoof and a short hoof.

Since I reside in California, Carol Chapman of Boerne, Texas, graciously agreed to grow long feet on two of her Arabian horses for the tests. I flew to San Antonio and Carol arranged motel and transportation to Texas A & M. Carol and her friend Tom Wood picked me up at the motel at 6:00 a.m. on June 27 and we hauled the two stallions with long feet to the University to a 10:00 a.m. schedule. It was a hot, humid day and the tests took a good 6 hours to complete.

Not knowing in advance what to expect, Carol could not really prepare the horses ahead of time. However, these two stallions were wonderfully willing to walk, normal trot, and extend their trot down a long, narrow strip of black rubber matting over the force plate after only a few trial runs--a testimony to the wonderful Arabian disposition and intelligence! Dr. Auer and staff were efficient, courteous, and just plain terrific hosts. We all left the test very tired, hot, and sweaty, but feeling confident about the purpose for being there.

Our first step proved short hoofs and proper angle are a MUST in the maintenance of sound, healthy horses. It also proved how easily we can alter the balance and weight distribution away from the areas structured throughout the entire body to support peak forces and strains by alterations in the hoof length and angle, thus leaving the horse very vulnerable to injury and/or improper movement. We MUST conform our ideas of showing, riding, gaiting, etc., to the way the horse is structured. We cannot force the horse into our wrongly-conceived ideas of how long his hoof should be NO MATTER WHAT WE HAVE TAUGHT OUR MINDS TO ACCEPT AS WHAT WE WANT TO SEE, OR WHAT WE THINK LOOKS GREAT, OR WHAT WE THINK IS EXCITING TO SEE OR RIDE! The Arabian's true movement and proper appearance are dictated by the parameters within HIS STRUCTURE! Again, NATURAL is the key word for healthy soundness and preserving the true movement and appearance of the Bedouins' Arabian horse. Other breeds should also conform in movement to each breed's general over-all structure with each individual horse dictating individual differences within that breed. No fancy training or farrier work can change the STRUCTURE of any horse! They usually only create pressure points and strain in the wrong places to which the body objects.

For those of you who have the idea that a big horse needs longer feet (this may be a more prevalent misconception within the Arabian breed), I must relate from personal experience to the contrary. The largest Arabian I ever owned since 1960 though today was a 15-3 hand mare weighing 1255 pounds, who wore a size 2 shoe (American size). That's a big foot! She showed halter, costume, English pleasure and park with an occasional western class thrown in. She was ridden in the mountains over rugged trails of the wilderness areas such as the middle fork of the Kern River in the Sierra Nevada Range in California. She ran cattle up on those same mountains, and she was ridden for pleasure many, many miles in the Mojave desert from Point "A" to Point "B." She never carried more than a 3 1/2" toe (not counting the shoe). She wore a size 2 shoe--Diamond brand--which weighs 12 oz. I never changed foot length or shoeing on this mare no matter what I did with her. She was shod to her natural length and angle. She never took a lame step for the 15 years I owned and rode her.

All my horses are trimmed or shod every 6 to 8 weeks, depending on each individual horse. For over 50 years I have ridden horses from 14 hands to 16+ hands. These horses were trimmed from a 3" to a 3 1/2" toe length. These horses were ridden from western pleasure to stock horses; from English pleasure to park horses; from endurance and trail riding to working cattle in areas of sandy desert to harsh, rocky canyons and up steep, high Sierra Nevada mountain trails (in places these trails are blasted out of solid granite and left very ragged). Many times the same horse performed all of the above (very well, I might add). I have never had one of these horses go lame. I can guarantee, most of this riding and footing was far more punishing than any hard-packed show ring of today. Yet, as previously emphasized, we have a terrible percentage of our show horses today that claim to need "bute," pads, etc., to stay sound in order to make it through a class or warm-up session! There is a definite message here!

However, I must caution extreme care be used here! A short foot CANNOT have the sole, bars and frog cut out as done by many of today's farriers. The tried and true Cavalry method of leaving the "guts" of the foot intact MUST BE FOLLOWED! This is another whole subject better left for another time, but I feel it is imperative to direct your attention to this commonly-made error. I have seen horses not only lamed, but also foundered from "cutting the floor out of the foot." By the same token, I have also seen big horses with a foot too small for their size and weight which has been caused from farriers, over a period of time, cutting out the sole, bars, and frog on long hoofs which leaves no support to hold the walls out and the entire foot contracts. Small feet can, of course, be a result of genetic inheritance as happens in “some” breeds, especially when light-horse breeds are crossed on draft breeds. However, I believe most of the small feet we see in today's show horses are caused by man's lack of knowledge of proper foot care.

I am publishing this work in book form to share with the general public the progress that has been made so far which, although initiated because of problems in the Arabian horse breed, also applies to all horses. From here, the direction is UP! This has turned out to be an overwhelming job for one person and I encourage the younger generations to continue the fund raising and forward momentum of this project. I was 25 years younger when I organized this particular test. I have kept up the constant communications of what is right for our Arabians, but must now move to documentation of my work instead of live pursuit.

At this time, I wish to extend a very special thanks to the following organizations and individuals who believed in my goals and proposed procedures and who donated the funds to pay for the first step of this project.

1. Region 2, IAHA, Central California
2. Association of Ridgecrest Arabian Breeders (ARAB), Ridgecrest and Inyokern, California
3. Marylin Cerniga's Registered Short Horn Cattle, Hidden Valley Farm, Springville, California
4. Earle and Frances Hurlbutt, Santa Clarita, California
5. Ms. Karen Joyner, Sage Flat Arabian, Olancha, California
6. San Joaquin Valley Arabian Horse Association, Region 2, California
7. Sierra Dawn Arabians, Inyokern, California

My personal thanks to Carol Chapman for riding lap after lap over the force plate and to both Carol and Tom Wood (Tom rode a couple laps, too) whose cooperation in providing horses with long feet and whose help with transportation and all the details made the difficulties and problems of coordinating this project from California so much easier and more successful.

September, 2013

Influence of Foot Length on the Gait of Arabian Horses

The following tests were conducted on June 27, 1988, at Texas A & M University using the Kaegi Gaitanalysis, force-plate system under the direction of Dr. J. A. Auer, American College of Veterinary Surgeons and Professor of Large Animal Surgery. This first section is by Lorry Wagner and is an edited (or simplified) version of Dr. Auer's final report on the full results of this project for easier understanding of the over-all testing results. A full, unedited copy of Dr. Auer's analysis plus all figures and tables produced by the force-plate follows this section.

The Kaegi Gaitanalysis system is a force-plate system consisting of many, many sensors laid down and covered with a sensor carpet so a horse may travel over the sensors. The sensors record the force expelled over each one as the horse's weight moves forward over that hoof. These readings are automatically recorded on a computer system which is housed adjacently in a temperature-controlled room. Peak force per sensor is expressed in pounds per square inch (psi). This pressure is derived from the weight placed on the sensor by the horse, minus the intrinsic pressure within the sensors. This peak force, therefore, does not comprise the absolute force placed on the sensors. Nevertheless, when making comparison measurements, it does indicate the correct differences in peak force.

Horse No. 1, 16-year-old Arabian Stallion (Que Natta)
Weight: 1,000 pounds
Height: 14-3
Starting Hoof Length, Front Feet: 5 1/4" (without shoe)

Horse No. 2, 6 year-old Arabian Stallion (Wind of Fire)
Weight: 1,085 pounds
Height: 14-2 1/2
Starting Hoof Length, Front Feet: 5 1/4" (without shoe)

Both horses had been left barefooted behind so had retained somewhat normal length of rear hoofs from natural wear. The hind feet were trimmed along with the front feet, but no significant change of length occurred.

Each horse was ridden over the sensor carpet at the walk, normal trot and extended trot with the long hoof. Both horses were then trimmed to a toe length of 3 1/2" (without shoe) with an angle of 54 degrees on Horse No. 1 and an angle of 55 degrees for Horse No. 2. Both these horses traveled in the more natural stride of the Arabian. In addition to the force plate, high-speed videography also recorded movement.

Following are the statistical results:

PEAK FORCE

Measured in pounds per square inch concussion/location of sensor under hoof where peak force registered.*

	Left Front	Right Front	Left Hind	Right Hind
Table No. 1-A				
Horse No. 1 (Que Natta)				
Long Hoof	**7.36 psi/2.4***	**7.40 psi/2.6**	**8.34 psi/6.6**	**8.48 psi/5.6**
Short Hoof	**6.96 psi/5.8**	**6.97 psi/4.6**	**9.02 psi/4.2**	**9.04 psi/5.6**
Table No. 2-A				
Horse No. 2 (Wind of Fire)				
Long Hoof	**8.44 psi/2.5**	**7.77 psi/2.3**	**6.98 psi/5.8**	**6.98 psi/5.3**
Short Hoof	**7.95 psi/5.5**	**7.32 psi/5.8**	**8.32 psi/5.5**	**8.25 psi/5.4**

Each hoof covers several sensors. For each foot the sensors on the force plate are numbered starting with number 1 at the heel, number 2 next forward, etc., continuing to the toe. The numbers here following the slash are the sensor numbers where peak force of concussion was recorded. These numbers are an average of the runs. See Table I on the next page for each individual run.

Keeping in mind that the front feet only were long, notice the force was recorded under the heel in each case until the hoof was shortened when it then moved forward toward the toe where the structure of the foot was designed to bear and deal with this peak force. In addition, note that after trimming the rear feet bore significantly more weight and the weight in front was decreased as it should be. Table I shows each run individually plus tabulates the "averages of each foot as depicted in the above graph.

TABLE I
Horse No. 1 (Que Natta)
Long Foot--5 1/4 inches (without shoe)

	Time (MSec)				Force (Psi/Sensor)*			
	FL	HL	FR	HR	FL	HL	FR	HR
Run 6	272	308	281	253	8.3/2*	7.8/7	7.0/3	8.4/4
	287	339	266	296	7.2/3	10.2/6	7.5/3	13.0/6
Run 7	289	231	270	242	7.0/3	8.2/6	7.7/2	7.5/6
	-	-	384	293			7.7/3	7.3/1
Run 8	264	289	287	244	6.7/2	8.0/7	7.2.8	6.0/5
	262		257	267	7.6/2	7.5/7	6.8/2	8.7/7
Average	274.8	292.0	272.8	265.83	7.36	8.34	7.40	8.48
SEM**	5.65	17.60	5.34	9.76	0.28	0.48	0.20	0.98

Trimmed and Reset--3 1/2 inches (without shoe)

	FL	HL	FR	HR	FL	HL	FR	HR
Run 16	245	246	243	297	6.5/5	9.5/5	6.3/4	10.7/5
	270	220	-	-	7.8/6	8.8/5	-	-
Run 17	256	263	235	230	7.5/6	10.0/4	7.0/5	9.3/7
	-	-	243	285	-	-	7.3/5	10.3/6
Run 18	277	237	273	214	6.0/6	9.0/3	7.2.5	8.0/5
	-	-	250	294	-	-	7.5/6	8.0/6
Run 19	250	262	238	260	7.0/6	7.9/4	6.2/4	8.5/5
	-	-	251	314	-	-	6.0/3	8.5/5
Average	259.6	245.6	247.6	263.3	6.96	9.02	6.97	9.04
SEM**	6.04	8.07	4.77	14.26	0.33	0.37	0.22	0.41

NOTE:

* The Sensors on the force plate are numbered starting with number 1 at the heel, number 2 next forward, etc., continuing to the toe. The numbers above following the slash are the sensor numbers where peak force of concussion was recorded. Keeping in mind that the front feet only were long (the hind were normal length) on the first runs 6, 7 and 8 (front were trimmed from 5 1/4" to 3 1/4"--without shoe--hind were barefoot and approximately 3 1/2" to start and trimming was insignificant) and shortened on second runs 16, 17, 18, and 19 as just described, the significant shift of peak force toward the toe where the structure of the foot was designed to bear and deal with this peak force is a critically important statistic. This statistic demands shorter hoofs for maintaining the health and soundness of our Arabians.

**Standard Error of the Mean

LENGTH OF TIME HOOF WAS ON THE GROUND

Measured in milliseconds (these figures are the average of the individual runs show on Table II, next page).

	Left Front	Right Front	Left Hind	Right Hind
Table No. 1-B				
Horse No. 1 (Que Natta)				
Long Hoof	**274.8**	**272.8**	**292.0**	**265.8**
Short Hoof	**259.6**	**247.6**	**245.6**	**263.3**
Table No. 2-B				
Horse No. 2 (Wind of Fired)				
Long Hoof	**387.0**	**371.6**	**303.4**	**344.7**
Short Hoof	**328.5**	**321.7**	**288.5**	**329.3**

TABLE II
Horse No. 2 (Wind of Fire)
Long Foot--5 1/4 inches (without shoe)

	Time (MSec)				Force (Psi/Sensor) *			
	FL	HL	FR	HR	FL	HL	FR	HR
Run 11	411	335	421	354	8.2/3*	6.5/5	7.5/3	7.6/5
	-	-	434	383	-	-	8.3/2	7.6/5
Run 12	381	256	308	367	9.0/2	8.2/6	8.2/2	6.8/6
	-	-	425	-	-	-	8.3/2	-
Run 13	411	324	333	311	9.0/2	6.7/6	7 .5/2	6.5/5
	-	-	373	295	-	-	7.3/2	6.9/5
Run 14	385	317	307	358	7.8/2	6.5/6	7.3/3	6.5/6
	387	285	-	-	8.2/3	7.0/6	-	-
Average	387.0	303.4	371.6	344.7	8.44	6.98	7.77	6.98
SEM**	12.15	14.5	21.21	13.94	0.24	0.32	0.18	0.21
Trimmed and Reset--3 1/2 inches (without shoe)								
Run 22	335	248	290	325	9.0/5	9.0/5	8.0/6	8.0/5
	380	286	-	-	9.3/5	7.5/5	-	-
Run 23	307	316	281	324	9.0/6	9.4/5	7.1/5	7.2/5
							7.01/6	8.0/4
	-	-	411	334	-	-	7.5/6	10.1/5
Run 24	248	292	363	317	7.8/5	9.3/6	8.0/6	7.0/5
	-	-	302	337	-	-	7.0/6	8.1/7
Run 25	289	300	383	339	7.6/6	7.5/6	6.3/6	8.1/7
	412	289	-	-	7.0/6	7.2/6	-	-
Average	328.5	288.5	321.7	329.3	7.95	8.32	7.32	8.25
SEM**	25.8	9.22	24.67	3.53	0.35	0.42	0.27	0.40

NOTE:

***The Sensors on the force plate are numbered starting with number 1 at the heel, number 2 next forward, etc., continuing to the toe. The numbers above following the slash are the sensor numbers where peak force of concussion was recorded. Keeping in mind that the front feet only were long (the hind were normal length) on the first runs Nos. 11, 12, 13 and 14 (front were trimmed from 5 1/4" to 3 1/4"--without shoe--hind were barefoot and approximately 3 1/2" to start and trimming was insignificant) and shortened on second runs Nos. 22, 23, 24 and 25 as just described, the significant shift of peak force toward the toe where the structure of the foot was designed to bear and deal with this peak force is a critically important statistic. This statistic demands shorter hoofs for maintaining the health and soundness of our Arabians.**

****Standard Error of the Mean**

CONCLUSIONS

1. Long Hoof

a. Greater concussion occurs on front feet and limbs.
b. Peak force of this concussion was directly at the heel on the front feet, predisposing the horse to injury in the tendons, suspensory apparatus, as well as the front part of the fetlock joint. Increased compression forces on the front part of the forelimb can result in chip fractures. In addition, the long feet also tend to show more hoof problems such as quarter cracks, scalping, and other injuries to that region.
c. The length of time the foot was on the ground bearing the horse's weight was longer (previous pages). This means the horse's structure was bearing weight a longer time with each stride; i.e., more wear, tear, and fatigue of bone, muscle, tendons, etc., engaged in supporting the horse.
d. Videography (Figures 1, 2, 6 and 7 in Dr. Auer's following report showed uneven speed of movement in hoofs and also fetlocks.
e. Horse's center of gravity (balance) moved to being heavy on the forehand which causes increase of weight and concussion on the forehand--an extremely vital statistic in that this condition also predisposes the horse to injury and/or lameness such as road founder.
f. Videography showed the angle of the back at a greater angle, decreasing from the withers to the root of the tail.*

2. Short Hoof

a. Significantly less concussion on front feet and limbs, shifting more of the burden of weight bearing to the hindquarter which is designed for this purpose.
b. Peak force of much less concussion was moved significantly forward toward the toe where the hoof is designed to accept this force.
c. The length of time the foot was on the ground bearing the horse's weight was noticeably shorter, thus the horse's weight-carrying structure was subjected to use for less time with each stride. This translates into less wear, tear, and fatigue of these areas.
d. Videography showed more even speed of movement in hoofs and also in fetlocks.
e. Horses center of gravity (balance) moved back to correctly distribute more weight toward the hindquarter.
f. Videography showed angle of the back became flatter.*

**Dr. Auer feels the angulation of the back as described in "1.f." above probably is a reflection of the length of front hoofs. However, keeping in mind the fact that the angle of the back became more level when the hoofs were shortened and more weight was shifted to the hindquarter ("2.f." above), Dr. Auer and I agree that more tests are indicated to substantiate or disprove the currently accepted theory by many that the forehand "raises" when a horse's center of gravity (balance) shifts more weight to the hindquarter. Dr. Auer believes the word "lighten" is the correct term; however, through translation and interpretation, this has been described many times as lifting or raising the forehand--we may have an error in terminology definition that over the years has created a problem in a number of riders and trainers trying to literally "raise" the forehand.*

The obvious over-all conclusion is that the longer the hoof, the more the horse is predisposed to many kinds of injury to the forehand cause by 1) peak force of concussion being directly under the heel, 2) interference with the natural balance of the horse by too much weight- bearing on the forehand, and 3) the hoof stays on the ground longer bearing peak force and over-all weight a greater period of time with each stride; i.e., weight distribution and peak forces not being correctly distributed over the structures of the horse which are meant to carry and handle that weight exposes the horse to lameness and/or physical problems higher in the body..

The short hoof results in 1) the area receiving the peak force of concussion moving forward to the area of the hoof designed to support this force, 2) the horse's center of gravity (balance) moves back toward the hindquarter lessening the concussion on the forehand, and 3) the hoof spends less time on the ground; i.e., weight-bearing and concussion on correct areas and less stress and fatigue of these weight-bearing structures are in coordination with the horse's natural structure. These tests also show that length of foot had no significant influence on the length of the stride. Length of stride tended to LENGTHEN with the SHORTER hoof; however, not to any great degree.

The test results reported herein are only the first step. It then followed that I must find a way to communicate the importance of retaining the natural Arabian movement with all people involved in Arabians. Thus, this report giving the test results was published in an Arabian breed magazine in 1989. I re-emphasize that improper care can lead to lameness and other problems in many places over the body, as the 4 graphs in Dr. Auer's following report show by high-speed videography how the entire body carriage is affected by long versus short feet which also shift the peak concussion point within the foot structure.

These test results have brought forth spirited debate within the Arabian horse show-ring enthusiasts. My very strong position first raises the question, "Whatever happened to the natural gaits of the original Arabian horse in our shows?" It seems over the years most of our breeds' show-ring competitions have led to parallel problems; however, I specifically address the Arabian breed within these pages since the Arabian has unique attributes different from other breeds. Since all breeds and even grade horses for that matter must have proper foot care focused on making sure the naturalness of movement of that particular horse is not compromised, I bring the originally published work up to date and share it in this book with the general public. Those interested in other breeds will find the parallels for their breed very easily within the this over-view of what happened within the Arabian breed and how these test results are not breed specific, but do apply to all horses.

Sierra Dunes and author.

Pictured at a show between classes at the age of 10, "Dunsey" was a finished, bridled horse and won Western Pleasure, Stock Horse, English Pleasure, Park Horse and Arabian Costume this day. He was the perfect example of a well-trained, extremely versatile Arabian stallion.

"Dunsey" received foundation training in the snaffle, moved on into the hackamore, was bridled into the half breed and finished in the Spanish spade. The author and horse were so finely tuned to each other that one could not see the signals between them

He was always trimmed and shod with short hoofs measuring a 3" toe length and never wore a shoe weighing more than 10 ounces. He lived until the age of 30 and was ridden to the end, just slowing down a bit from old age.

Report of a Preliminary Study on the Influence of Foot Length on the Gait in Arabian Horses

Dr. J. A. Auer,
Dept. of Large Animal Medicine & Surgery
Texas Veterinary Medical Center
College of Veterinary Medicine
Texas A&M University

Prepared August 16, 1988

Report of a Preliminary Study on the Influence of Foot length on the Gait in Arabian Horses

On June 27, 1988, two horses owned by Ms. Carol Chapman of Boerne, Texas, were evaluated at Texas A&M University, Large Animal Clinic.. During the months previous to the above date, Ms. Chapman allowed the feet of the horses to grow without being trimmed. The animals were then first evaluated with long feet and subsequently the feet were trimmed by an expert farrier, shoes reapplied and the horses re-evaluated. The evaluation consisted of force analysis (kinetic evaluation) over the Kaegi Gaitanalysis system and high speed cinematography evaluation using a high shutter speed VHS camera and the Peak Performance Technology software analysis, which was derived through digitation of various body points during the trot.

Horse #1: A 16 year old Arabian Stallion by the name of Que Natta.

Weight: 1000 pounds without rider, 1155 pounds with rider and saddle. Foot length, 5 1/2”

After trimming, both front feet were set at an angle of 54 degrees and a foot length of 3 3/4” from coronary band to the ground (including the shoe).

I. Kinetic evaluation

Peak force per sensor is expressed in pounds per square inch. (This pressure is derived from the weight placed on the sensor by the horse minus the intrinsic pressure within the sensors.) This peak force therefore does not comprise the absolute force placed on the sensors. Nevertheless it is a good indicator for the changes seen. Before trimming, the peak pressures in the left and right front feet were 7.36 psi and 7.4 psi respectively; in the left rear leg: 8.34 psi, right rear: 8.48 psi (Table 1). This means that the animal placed more weight on the rear limbs than the forelimbs. After the shoeing, the pressures changed to the following parameters: 6.96 psi in the left front; 6.97 in the right front; 902, in the left rear; 904 in the right rear (Table 1); indicating that the pressure decreased in the forelimbs and increased weight was placed on the rear limbs after setting the foot at the proper angle and length. Also it was interesting to note that the peak pressure was located on the sensor over the heels before shoeing and was detected just behind the toe after they were trimmed (Table 1).

Que Natta (Long Feet).
1972 Arabian

TABLE 1

	TIME (MSec)				FORCE (Psi/Sensor)			
	FL	HL	FR	HR	FL	HL	FR	HR
Run 6	272	308	281	253	8.3/2	7.8/7	7.0/3	8.4/4
	287	339	266	296	7.2/3	10.2/6	7.5/3	13.0/6
Run 7	289	231	270	242	7.0/3	8.2/6	7.7/2	7.5/6
	-	-	384	293			7.7/3	7.3/1
Run 8	264	289	287	244	6.7/2	8.0/7	7.2/8	6.0/5
	262		257	267	7.6/2	7.5/7	6.8/2	8.7/7
X	274.8	292.0	272.80	265.83	7.36	8.34	7.40	8.48
M	5.65	17.60	5.34	9.76	0.28	0.48	0.20	0.98
TRIMMED and RESET								
Run 16	245	246	243	297	6.5/5	9.5/5	6.3/4	10.7/5
	270	220	-	-	7.8/6	8.8/5	-	-
Run 17	256	263	235	230	7.5/6	10.0/4	7.0/5	9.3/7
	-	-	243	285	-	-	7.3/5	10.3/6
Run 18	277	237	273	214	6.0/6	9.0/3	7.2/5	8.05
	-	-	250	294	-	-	7.5/6	8.0/6
Run 19	250	262	238	260	7.0/6	7.9/4	6.2/4	8.5/5
	-	-	251	314	-	-	6.0/3	8.5/5
X	259.60	245.60	247.57	263.33	6.96	9.02	6.97	9.04
SEM	6.04	8.07	4.77	14.26	0.33	0.37	0.22	0.41

NOTE:

***The Sensors on the force plate are numbered starting with number 1 at the heel, number 2 next forward, etc., continuing to the toe. The numbers above following the slash are the sensor numbers where peak force of concussion was recorded. Keeping in mind that the front feet only were long (the hind were normal length) on the first run (front were trimmed from 5 1/4" to 3 1/4"--without shoe--hind were barefoot and approximately 3 1/2" to start and trimming was insignificant) and shortened on second run as just described, the significant shift of peak force toward the toe where the structure of the foot was designed to bear and deal with this peak force is a critically important statistic. This statistic demands shorter hoofs for maintaining the health and soundness of our Arabians.**

****Standard Error of the Mean**

The support time of each foot (the time each foot stayed on the centers) was changed insignificantly between the front and the rear feet before trimming as well as after trimming (Table 2). There was a tendency that the feet stayed on the ground less time after the feet were trimmed.

Que Natta 1972 Arabian Stallion TABLE 2

Step Width (mm)

LONG TOES

File #	FL - FR	FR - FL	FL - HR	HR - HL	HL - HR	HR - FL
Run 6	1075 1050	975	1100	975 1025	1100	1000 1050
Run 7	1125	1175	1150	1225	1000	1100
Run 8	1100	1075	1075	1125	1075	1100
		1100		1100		1100
X	1031.25	1037.5	1031.25	1045.83	1040.	1066.67
SEM	62.07	46.0	78.65	56.4	20.3	16.67
TRIMMED AND RESET						
Run 16	1000	1050	1000	1075	1000	1050
Run 17	1050	1075	1050	1125	1025	1075
Run 18	1050	1050	975	1075	975	1000
Run 19	1050	1025	1025	1025	1025	1000
X	1031.25	1050.0	1012.5	1075.0	1006.25	1031.25
SEM	12.0	10.2	16.15	20.4	12.0	18.75

NOTE:

***The Sensors on the force plate are numbered starting with number 1 at the heel, number 2 next forward, etc., continuing to the toe. The numbers above following the slash are the sensor numbers where peak force of concussion was recorded. Keeping in mind that the front feet only were long (the hind were normal length) on the first run (front were trimmed from 5 1/4" to 3 1/4"--without shoe--hind were barefoot and approximately 3 1/2" to start and trimming was insignificant) and shortened on second run as just described, the significant shift of peak force toward the toe where the structure of the foot was designed to bear and deal with this peak force is a critically important statistic. This statistic demands shorter hoofs for maintaining the health and soundness of our Arabians.**

****Standard Error of the Mean**

The step widths were evaluated and we found in most parameters no significant changes between the left and right front feet, the left and right rear feet as well as the two diagonals before the feet were trimmed. Interesting enough we found a significant difference in the stride length between left rear and right rear feet after trimming (Table 4).

Wind of Fire

TABLE 4

Step Width (mm).

Long Feet

File #	FL - FR	FR - FL	HL - HR	HR - HL	HL - FR	HR - FL
Run 11	1300	1300	1325	1300	1150	1100
Run 12	1350	1250	-	1375	1125	1150
Run 13	1275	1325	1325	1250	1175	1150
Run 14	1325	1350	1375	1325	1175	1150
X	1325.0	1306.25	1341.67	1312.5	1156.25	1137.5
SEM	17.7	21.35	16.67	25.0	12.0	12.5
TRIMMED AND RESET						
Run 22	1275	1275	1275	1200	1100	1100
Run 23	1300	1300	1275	1225	1125	1050
Run 24	1250	1300	1250	1225	1150	1125
Run 25	1300	1200	1250	1175	1150	1100
X	1281.25	1268.75	1262.5	1206.25	1131.25	1093.75
SEM	12.0	23.66	7.22	12.0	12.0	15.73

II. Kinematic Evaluation (high-speed videography)

The back line was at a greater angle, decreasing from the withers to the root of the tail, before trimming compared to after trimming. This is an obvious fact because we remove 1 1/2" from both front feet. Therefore this angle should be more flatter following trimming (Figures 1 & 2, Horse No 1 and Figures 6 & 7, Horse No. 2)..

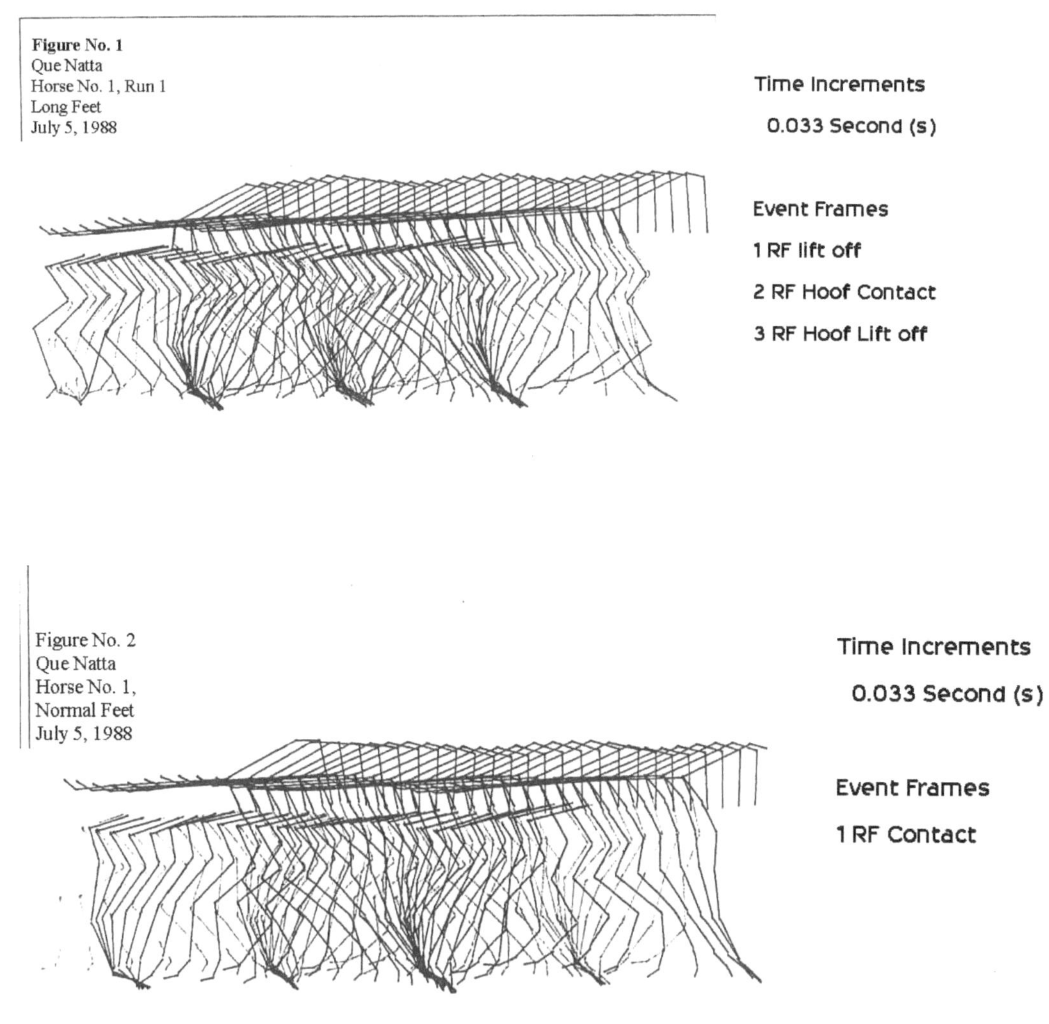

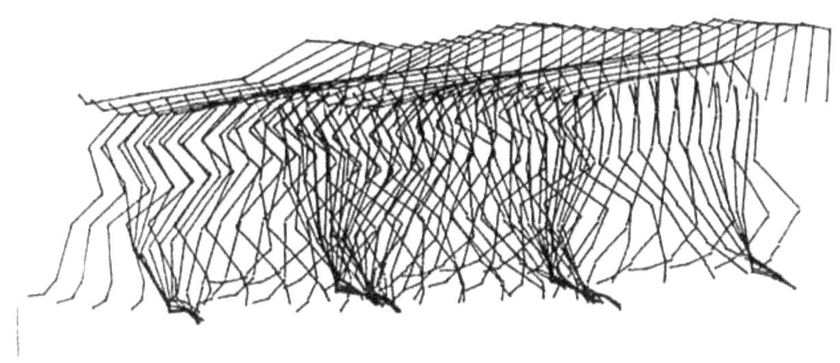

Wind of Fire
Horse No. 2
1982 Arabian
Stallion
long feet
Time Increment
0.033 second(s)
Event Frames
1 RF CONTACT
HORSE #2 RUN 1
5 JULY 1988 LONG FEET
Fig 6

Wind of Fire
Horse No. 2
1982 Arabian
Stallion
normal feet
Time Increment
0.033 second(s)
HORSE #2 RUN 2
5 JULY 1988 NORMAL FEET

Fig 6

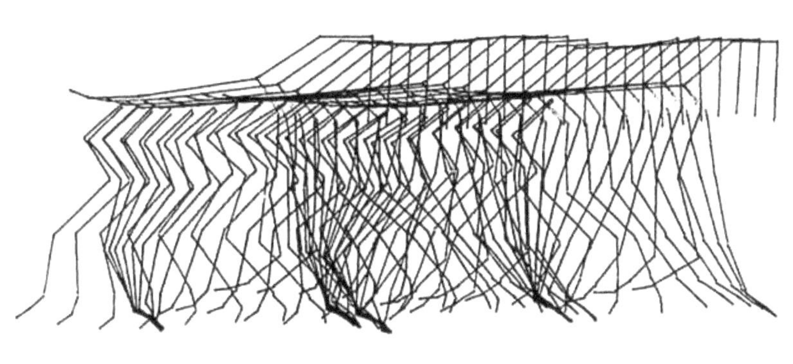

Fig 7

The speed at which the feet were moved during the trot is depicted on Figure 3 in the upper diagram, the left and right foot are compared before they were trimmed and in the lower diagram after trimming.

Figure No. 3
Que Natta
Horse No. 1, Runs 1& 2
Top, long hoof...Bottom short hoof
July 5, 1988

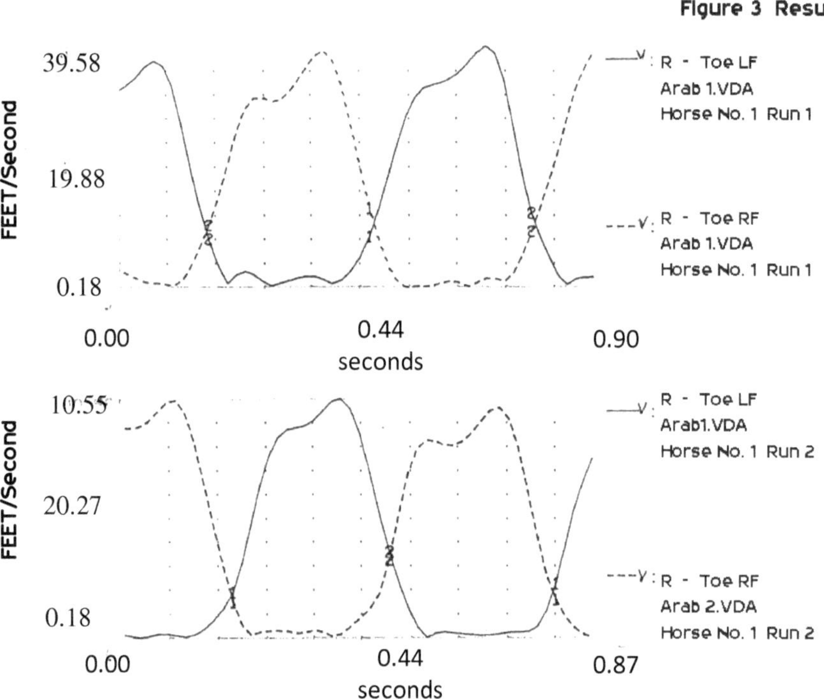

Top Diagram: Speed at which the feet were moved during the trot before trimming including shoe (5 1/2").

Lower Diagram: Speed at which the feet were moved during the trot after trimming including shoe (3 1/4")

No significant differences could be detected; however, both feet moved at more even speeds following trimming.

Acceleration of the feet demonstrating the speed with which the feet during the stride were accelerated and decelerated is shown in Figure 4. Again the upper diagram compares the acceleration of the left and right forefoot before trimming and the lower diagram after trimming . It is interesting to note that after trimming faster acceleration speeds were noted; however, the pictures are not uniform through the three steps.

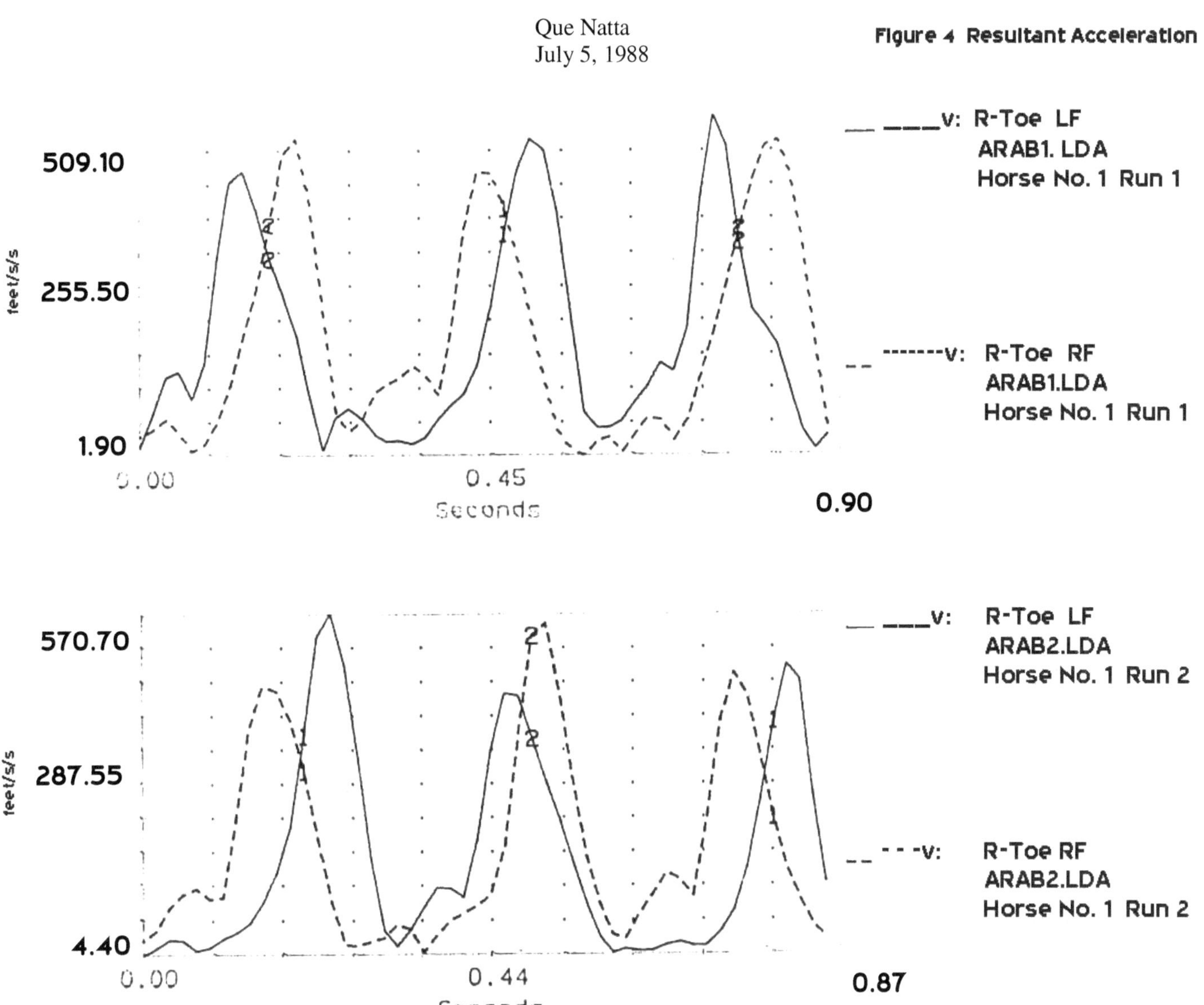

Top Diagram: Before trimming

Bottom Diagram: After trimming

Note that after trimming faster acceleration speeds were noted; however, the pictures are not uniform through the three steps.

The change of the fetlock angles during the stride are show in Figure 5, which demonstrates minor differences. However, normalizing the foot angles and lengths resulted in a more even movement of the two fetlocks.

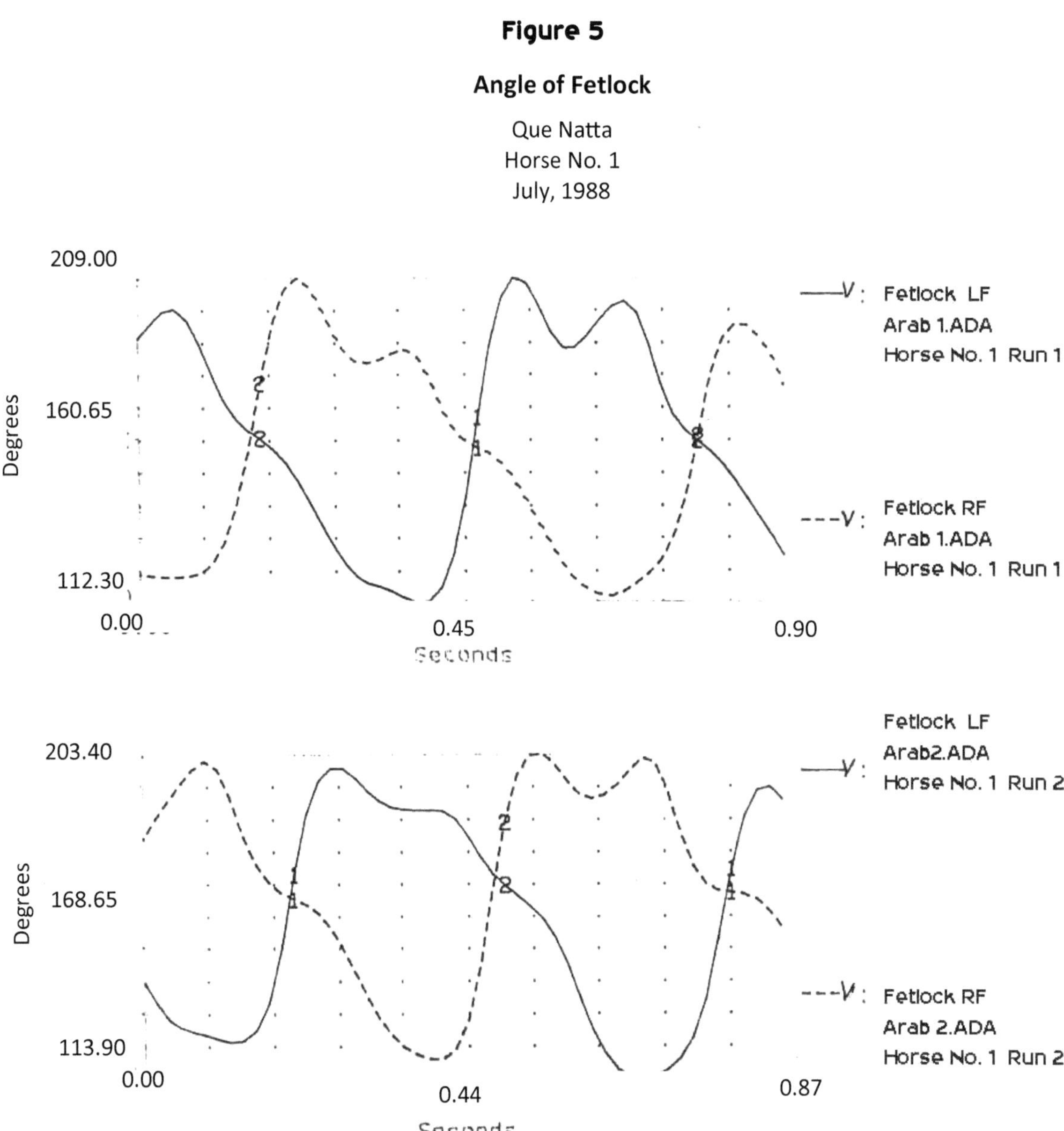

Fig 5

Horse #2: A 6 year old Arabian by the name of Wind of Fire.

Weight: 950 pounds without rider, 1085 pounds with rider and saddle. Foot length, 5 1/2"

I. Kinetic Evaluation

The Peak Force (pressure differential) before trimming was as follows: left front: 8.44 psi, right front: 7.77 psi, left rear: 6.98 psi, and right rear: 6.98psi. After trimming the following pressures were recorded: left front: 7.95 psi, right front: 7.32 psi, left rear: 8.32psi, and right rear: 8.25 psi (Table 3)

Wind of Fire 1982 Arabian Stallion — TABLE 3

File	TIME (Msec)				FORCE (Psi/Sensor).			
	FL	HL	FR	HR	FL	HL	FR	HR
Long Feet								
Run 11	411	335	421	354	8.2/3	6.5/5	7.5/3	7.6/5
	-	-	434	383	-	-	8.3/2	7.6/5
Run 12	381	256	308	367	9.0/2	8.2/6	8.2/2	6.8/6
	-	-	425	-	-	-	8.3/2	-
Run 13	411	324	333	311	9.0/2	6.7/6	7.5/2	6.5/5
	-	-	373	295	-	-	7.3/2	6.9/5
Run 14	345	317	307	358	7.8/2	6.5/6	7.3/3	6.5/6
	387	285	-	-	8.2/3	7.0/6	-	-
X	387.0	303.4	371.57	344.67	8.44	6.98	7.77	6.98
SEM	12.15	14.5	21.21	13.94	0.24	0.32	0.18	0.21
TRIMMED AND RESET								
Run 22	335	248	290	325	9.0/5	9.0/5	8.0/6	8.0/5
	380	286	-	-	9.3/5	7.5/5	-	-
Run 23	307	316	281	324	9.0/6	9.4/5	7.1/5	7.2/5
							7.01/6	8.0/4
	-	-	411	334	-	-	7.5/6	10.1/5
Run 24	248	292	263	317	7.8/5	9.3/6	8.0/6	7.0/5
	-	-	302	337	-	-	7.0/6	8.1/7
Run 25	289	300	383	339	7.6/6	7.5/6	6.3/6	8.1/7
	412	289	-	-	7.0/6	7.2/6	-	-
X	328.5	288.5	321.67	329.3	7.95	8.32	7.32	8.25
SEM	25.8	9.22	24.67	3.53	0.35	0.42	0.27	.40

This horse increased the weight bearing on the forefeet before the feet were trimmed. After trimming, the rear feet bore significantly more weight and the weight in the front feet decreased. The same tendency was noticed in the other horse. The top diagram shows the movement before trimming and the lower diagram shows it after the trimming. There is a great difference between the two diagrams. The lower diagram showed the two feet moving much more evenly compared to before.

The acceleration of the toe is shown in Figure 9.

Figure 9

Resultant Acceleration

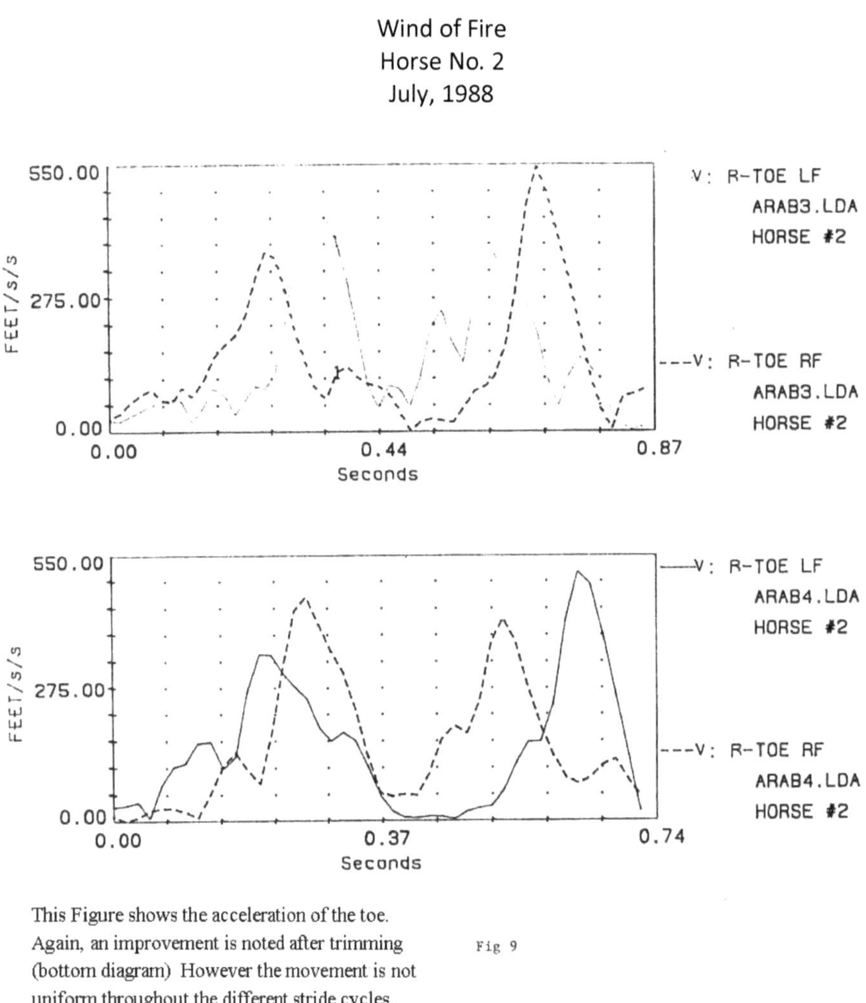

This Figure shows the acceleration of the toe. Again, an improvement is noted after trimming (bottom diagram) However the movement is not uniform throughout the different stride cycles.

Fig 9

Again, an improvement is noted after trimming (bottom diagram). However, the movement is not uniform throughout the different stride cycles.

Fetlock motion is shown in Figure 10 which shows little change between the before and after trimming.

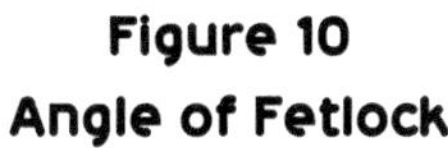

Wind of Fire
Horse No. 2
July, 1988

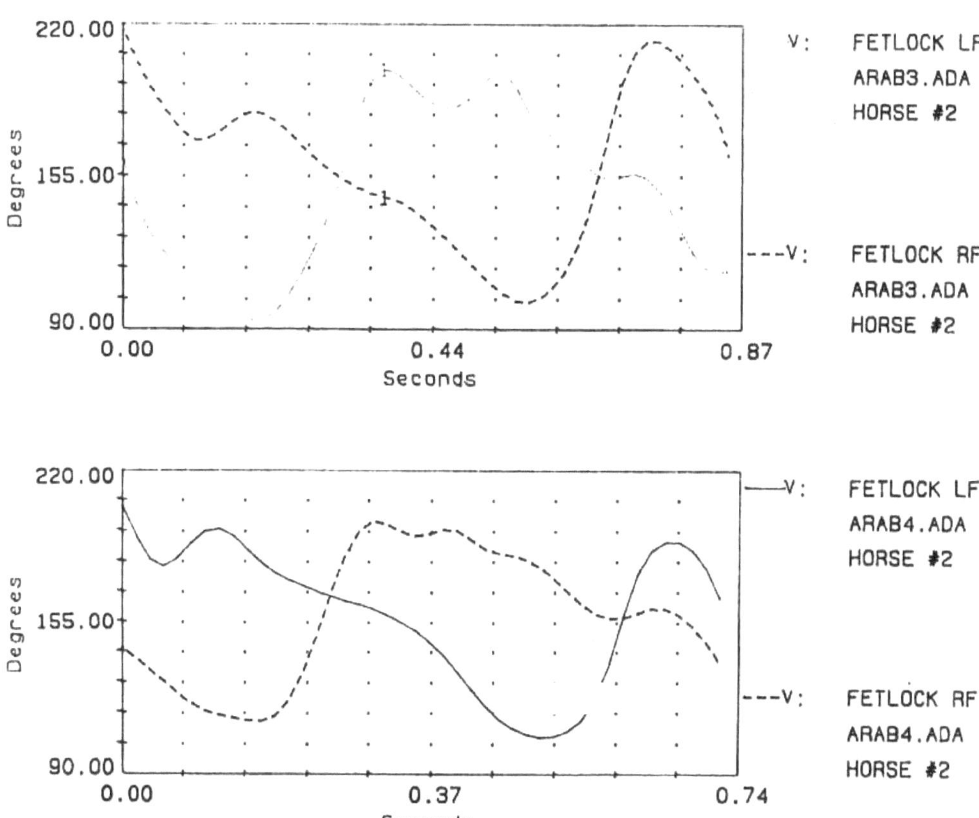

Fetlock motion is show in the above diagrams which show little change between before and after trimming.

Fig 10

Angle of Fetlock

Discussion

In this analysis of two horses each were analyzed and various differences and some inconsistencies in the gait before and after trimming were noticed. The feet were extremely long and the shoes overgrown before trimming. This is not a criticism; however, it explains some of the findings. We found a trend that with a shorter foot the weight shifted more on to the rear limbs. This is a trend which can be looked at as positive. The action in both front feet is not significantly difference in long feet versus rear feet. It was found that when the long feet were placed on the ground, most of the weight was placed in the heel region of both feet. This induces an increased tension in the tendon region as well as the suspensory apparatus (suspensory ligament and sesamoid bones). On the front part of the fore limbs increased compression forces are encountered, resulting in chip fractures. We found that once the feet were trimmed to a normal angle and balanced, the peak force was placed more towards the toe where it should be. Trimming also decreased the total amount of peak forces of the foot which is a beneficial aspect.

There are obviously differences found between the two horses and this has to be expected. By increasing the number of individuals tested, a more significant trend can be shown. Nevertheless, I am very positive that there is a difference in the font feet motion as well as weight bearing with long feet versus the short normal feet. Due to the facts encountered in this brief study, long feet are undesirable for the show horse because they increase the vulnerability of those animals to injury in the tendons, suspensory apparatus as well as in the front part of the fetlock joint. In addition to that, the long feet also tend to show more hoof problems such as quarter cracks, scalping and other injuries to that region. Some of those injuries occur due to the delayed break over at the toe. To improve motion even more the horses should be allowed to go barefoot. It could also be beneficial **in addition to the normal and normal length** to allow some horses half round shoes or rockered toe shoes or square toed shoes. All of these aspects decrease the stress and train on the tendons in the feet and facilitate early break over.

Respectively Submitted

J.A. Auer, Dr. Med. Vet, MS
Professor, VLAM

JAA/agm

www.ingramcontent.com/pod-product-compliance
Lightning Source LLC
LaVergne TN
LVHW070155110826
845147LV00002B/406
* 9 7 8 0 5 7 8 1 3 0 3 3 0 *